The Monkland Canal
Coal, Iron and Cold Hard Cash
Guthrie Hutton

The boatyard at the top of the Blackhill Locks is the likely location of this picture from about 1905 entitled *Passenger Motor Boat on Canal, Riddrie*.

Construction of the M8 Motorway between Glasgow Townhead and Easterhouse destroyed the canal, but replicated its route. New road or pedestrian bridges were built across the motorway to replace canal bridges.

© Guthrie Hutton, 2015
First published in the United Kingdom, 2015,
by Stenlake Publishing Ltd.
www.stenlake.co.uk
ISBN 9781840337211

The publishers regret that they cannot supply copies of any pictures featured in this book.

Printed by Blissetts, Shield Drive,
West Cross Industrial Park, Brentford, TW8 9EX

Acknowledgements

This little book is a reworking of my earlier volume *Monkland, the Canal That Made Money* published in 1993. Some pictures from that book have been included, along with a few from other books and a number of previously unpublished images. Pictures of the canal, especially between Glasgow and Drumpellier, are few, as if commercial photographers and well-to-do camera owners shunned what they probably regarded as a fading industrial relic. I am therefore grateful to Summerlee, Museum of Scottish Industrial Life, Culture NL Ltd., for permission to include the pictures on pages 14 and 36 and to East Dunbartonshire Archives and Local Studies for allowing me to reuse the images on pages 39 and 41.

Further Reading

Brotchie, T.C.F., *Some Sylvan Scenes Near Glasgow*, 1910.
Company History, *Stewarts and Lloyds Limited*, 1903-1953, 1953.
Drummond, Peter and Smith, James, *Coatbridge, Three Centuries of Change*, 1984.
Duckham, Baron F., *A History of the Scottish Coal Industry*, Vol. 1, 1970.
Geological Survey of Scotland, *The Economic Geology of the Central Coalfield of Scotland*, 1916.
Horsey, Miles, *Tenements and Towers*, 1990.
Lindsay, Jean, *The Canals of Scotland*, 1968.
Martin, Don, *The Garnkirk & Glasgow Railway*, 1981.
Martin, Don, *The Monkland & Kirkintilloch and Associated Railways*, 1995.
Miller, Thomas Ronald, *The Monkland Tradition*, 1958.
National Coal Board, *A Short History of the Scottish Coal-Mining Industry*, 1958.
Peden, Allan, *The Monklands, An Illustrated Architectural Guide*, 1992.
Thomson, George, *The Monkland Canal*, 1945 (reprinted).

Introduction

Glasgow wasn't flourishing. There was plenty of coal to heat its homes, but the men who owned the mines operated a cartel to limit supplies and keep prices high. Something had to be done and the idea that most appealed to the city magistrates was to get coal from the Monklands, but that needed a canal. An engineer named James Watt – who achieved lasting fame building steam engines – was asked to carry out a survey. He recommended two schemes and the magistrates unsurprisingly chose the cheapest, a level canal from Coatdyke to Germiston, near modern day Blackhill, in Glasgow. A subscription was opened, money was raised and Watt began work in 1770 at the eastern end.

From the start there were problems as ground conditions – hard clay soil in one place, running sand in another – defied the contractors, but Watt pushed them on. There were disputes too with Forth & Clyde Canal engineer, Robert MacKell, as each accused the other of stealing their most experienced navvies – navigators, men who made canal navigations. After three years the canal had not reached its intended terminal, but the money had been spent, and Watt was out of a job.

Within a few years much of Glasgow's tobacco-based wealth was lost along with Britain's American colonies and the unfinished canal became a liability. The few remaining proprietors managed to raise more money and in the early 1780s they extended the canal to Blackhill and made a lower level canal to Townhead in Glasgow. It still wasn't enough, but one principal backer, Andrew Stirling, assumed control along with his brothers John and James. They faced another problem; the Forth & Clyde Canal needed more water and could have ruined the Monkland by diverting its supplies. The two companies avoided that fate by acting together and, in a mutually beneficial scheme, joining their canals as one system. The link was completed with the opening of the Blackhill Locks in 1793.

Coal from the Monklands was now pouring in to Glasgow, mainly from Andrew Stirling's collieries; the city had achieved its objective. It got even better. In the 1830s the new hot-blast process of iron smelting was developed at Coatbridge and the canal's fortunes leapt. They continued to rise as vast quantities of coal and iron were sent to the city, and even the development of railways failed to stem the flow of commodities and cash. In 1846, at the height of the Monkland Canal's prosperity, it was taken over by the Forth & Clyde Canal Company and in 1867 the Caledonian Railway bought the two canals, and the money kept rolling in.

Trade began to decline around the 1880s. Inexorably, traffic drifted away until, by the mid 1930s, boat movements had ceased. The Caledonian Railway was taken over in 1923 by the new London, Midland and Scottish Railway, which tried to rid itself of the canal in 1942. That attempt failed, but formal abandonment was permitted in 1950 and the clamour to eliminate the weed-choked, rubbish-filled canal grew. Government funds paid for a pipeline to take the water through Coatbridge and to fill in the channel. Between Glasgow Townhead and Easterhouse the canal wasn't so much filled in as wiped from the map when the M8 motorway was built along its route. The most commercially successful canal in Scotland was thus consigned to history, with just a few remnants remaining as reminders of a once great enterprise.

A drawing made about 1910 of the canal, near Blairtummock.

The North Calder Water, seen here near Airdrie, was not an obvious source of water for the Monkland or Forth & Clyde Canals, but it became essential to both. Self-evidently, canals need water, but to the casual observer it is perhaps less obvious that the water flows, it is not stagnant, and so one of the main preoccupations of canal proprietors and engineers was to secure a constant supply. In parliamentary acts authorising the construction of canals, the rights to divert water into these man-made structures was enshrined in law, but for the proprietors of the Monkland Canal these provisions were barely adequate. Water could be drawn from sources three miles on either side of the canal, but that area overlapped the ten mile limit for the Forth & Clyde Canal and with the two canals being made at much the same time and both struggling for water, something had to give. The solution, to join the two canals and thus combine their water supplies was set out in an Act of Parliament in 1790, which also, specifically authorised drawing water from the North Calder.

At the time that the 1790 Act was being drafted, the Monkland Canal had terminals in the east at Sheepford (Coatdyke) and Glasgow Townhead in the west, but it was on two levels with a break at Blackhill. To make the plan work that gap had to be closed and extensions made, west to the Forth & Clyde at Port Dundas, and east to meet the North Calder 'at or near Woodhall or Faskine Miln'. This work was completed, and the two canals joined in 1793, but the Act also made provision for new reservoirs and in the late 1790s the Hillend Reservoir was made at Caldercruix. With a capacity of 750 million gallons Hillend was huge; two and a half times the combined volume of the Forth & Clyde Canal's main reservoirs, Birkenburn and Townhead. The water from Hillend didn't go directly into the canal; it was fed into the river so that a constant flow could be maintained, even in dry seasons, for other water users, like mills. A large weir at Woodhall directed water through a sluice into a narrow flume, which quickly widened out beside this cottage to form the main channel of the Monkland Canal: a modest start for an undertaking that was to have a major impact on 19th century Scotland.

With the original sections of the Monkland Canal having been made to get coal into Glasgow, it is scarcely surprising that, as soon as the eastern extension of the canal was completed, coal workings were developed along its banks. Principal amongst these was the colliery at Faskine opened on the north bank of the canal by landowner and Monkland Canal proprietor, Andrew Stirling. By 1825, eighteen collieries were sending coal to Glasgow along the canal, although some of these were situated at a distance from the canal and had to get their coal to it before it could be shipped. Woodhall became a trans-shipment point for coal brought by wagon-way across the North Calder Water from pits south of the river. By the mid 19th century, Faskine Colliery was owned and operated by William Baird & Co., who also worked a neighbouring pit at Palacecraig, but by that time much of the coal was being moved by rail. The remains of Faskine Colliery can be seen in the background of this picture of the bridge at Faskine East, or Upper Faskine, from the 1980s.

In their day canals were cutting-edge technology and attracted innovators like Thomas Wilson, the man who built Scotland's first iron boat, the *Vulcan* on the canal banks at Faskine. The exact spot is not known, but the location was evidently influenced by its proximity to ready supplies of coal and to the Calderbank Ironworks, famed for the quality of its malleable iron plates. These were fixed to a frame fashioned by blacksmiths John and Thomas Smellie, who hammered angle-iron ribs out of flat bars of iron. It must have been hard work, not helped by people who scoffed at the thought of floating iron, but the men who built her had the last laugh, because *Vulcan* didn't sink when launched in May 1819. She went into service as intended as a Forth & Clyde Canal passage (passenger) boat and was eventually scrapped in 1873. A replica of this historically important vessel was made for Monklands District Council as an exhibit at the Glasgow Garden Festival in 1988, where the picture was taken. Following the Festival she was taken to Summerlee Heritage Park, and in 2014, after being given a refit to make her suitable for display, was put on show at the revamped Summerlee, Museum of Scottish Industrial Life.

The words of the German writer, Goethe – *'the solution of every problem is another problem'* – neatly encapsulate the decision to abandon the canal. No longer used as a waterway, choked with weeds and unsightly with rubbish, something had to be done, but when the canal was closed it was still conveying water from the North Calder to the Forth & Clyde Canal and so, simply answering the siren call to 'fill it in' was not an option. It could have been restored and maintained as an attractive feature connecting Coatbridge with its hinterland, but instead it was drained and large diameter pipes were laid along the canal bed to carry the water. The pipeline was then buried; out of sight, but inevitably not out of mind, because the system still had to be monitored and maintained, to ensure that the water continued to flow. The Sikeside Weir, the eastern end of the pipeline through Coatbridge, is seen here in a picture from 1996. The canal remained in water to the east of the weir and the improved towpath alongside has been made into an attractive walkway. Regular weed control is needed, but this small remnant is a reminder of what might have been.

James Watt's contractors began digging the canal in 1770 at Sheepford, a site selected because of its proximity to the Airdrie South Burn and the water supplies that could be drawn from it. For the next twenty years this remained a dead end at the eastern end of the canal, but when the extension to the North Calder Water was made it suddenly became much more significant. The new section of canal was 21 feet higher than the old and so two locks, separated by an intermediate basin, were built. Known as the Sheepford Locks they are seen here about 1910 in a picture that shows the lower lock in the foreground, the intermediate basin and beyond that the bridge carrying the appropriately named Locks Street across the tail of the upper lock. The installation of locks meant that boats, previously restricted only by the width and height of bridges, also had to conform to the length of the locks. This did not affect the passenger boats because they only operated on the lock-free sections of water between Sheepford and Blackhill, and Blackhill and Townhead. In their hey-day, in the early 19th century before railways became popular, these boats could whisk their customers to or from the city in two and a half hours.

The basin between the two Sheepford Locks was also the junction for a branch canal, known as Dixon's Cut that meandered south to the Calder Iron Works on the banks of the North Calder Water. The works was established around 1800, but failed within three years, bankrupting its owners. The company supplying coal was left with unsold fuel and so its partners, one of whom was William Dixon, restyled themselves as the Calder Coal and Iron Company and bought the moribund works. They also linked up with one of its former associates, David Mushet whose discovery of locally abundant blackband ironstone revolutionised the Monklands iron industry. He left the area in 1805 triggering a break-up of the partnership and leaving Dixon in charge. A Northumbrian, he was still a teenager when he leased the Govan coalfield in Glasgow and began a career as a coal and iron entrepreneur. He eventually became sole owner at Govan and by acquiring other properties, including the Palacecraig and Faskine estates, his company became the second largest such concern in Scotland. Dixon's Cut was rendered obsolete by the railway age, but the Calder Iron Works, seen here in a picture from about 1905, continued in operation until 1921.

The works of A. & J. Stewart were situated just below Sheepford Locks on the north bank of the canal. Andrew Stewart had started out in business in 1860 with a tube-making works in St Enoch's Wynd in Glasgow, where the St Enoch's Centre was later sited. His brother James joined him as a business partner and in 1867 they opened their new Clyde Tube Works at Coatdyke. The buildings occupied about an acre of ground and a further two acres were available for expansion, which duly happened. The company also expanded; by 1882 it had become a private limited company and acquired the Sun Tube Works, also in Coatbridge, and the Clyde Pipe Foundry in Glasgow. Mergers and acquisitions in subsequent years brought other concerns into the company, including the Clydesdale Iron and Steel Company's Mossend works. The trend continued after Andrew Stewart's death in 1901 and in 1903 the company amalgamated with the Birmingham firm of Lloyd and Lloyd to form one of the giants of British heavy industry: Stewarts and Lloyds.

Stewarts and Lloyds were by far the biggest tube makers in Coatbridge with a number of sites including the Imperial, British, Clydeside and Sun as well as the Clyde works. Other companies, operating the Union, Victoria, Caledonian, Coatbridge and Coats works contributed to the town being the centre of tube making in Scotland. The collective capacity of these works in 1905 was such that over 1,800 miles of tubular products could be made in a month, including steam pipes, boiler tubes, refrigerator coils, water and gas mains, hydraulic tubes and tapered poles for electric tramways. Tube making was just one element of a Scottish malleable iron industry, dominated by Coatbridge, with many of the works, like the neighbouring Phoenix and Clifton works seen here, beside the Monkland Canal. From small beginnings in the 1830s the industry expanded through the middle decades of the 19th century until by 1900, in the wake of further modernisation, it could produce over 250,000 tons of finished iron a year. Even when the iron industry elsewhere was being ousted by mild steel, it was expanding in Coatbridge.

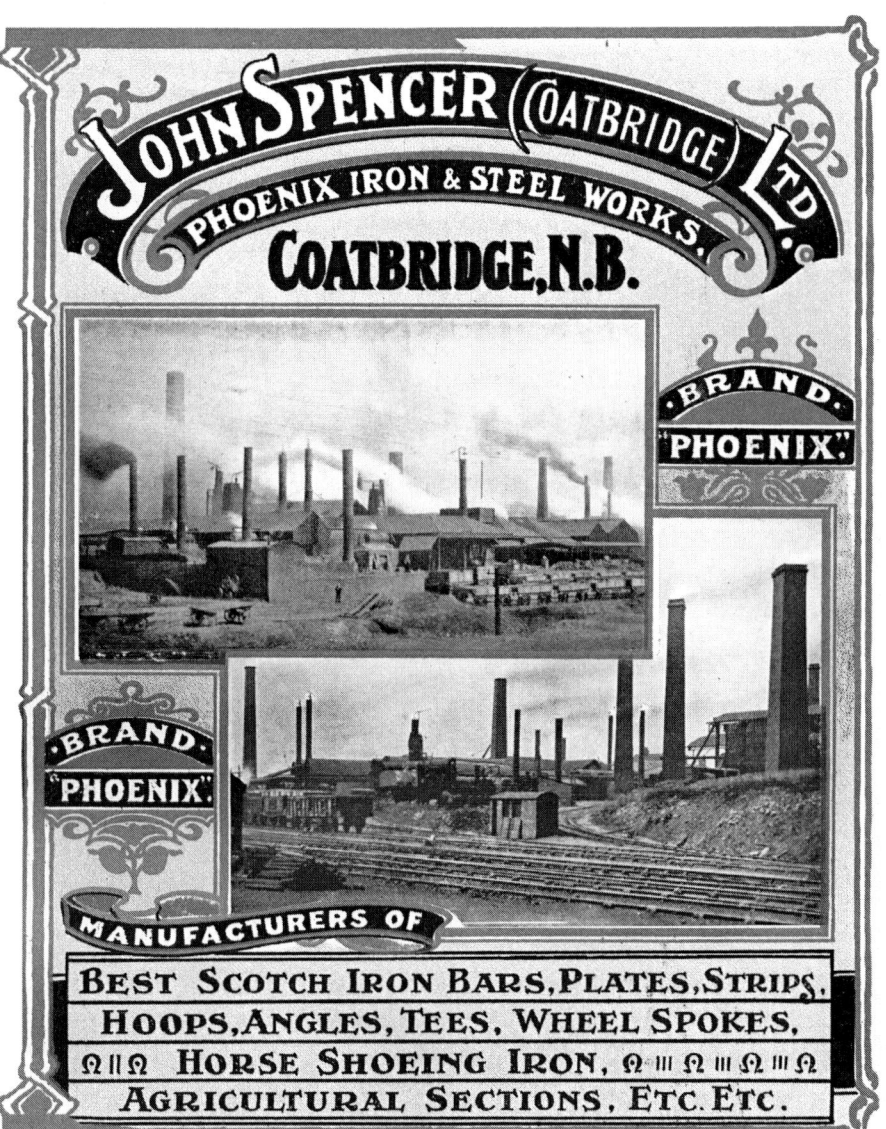

The Phoenix Iron Works, seen here in an advertisement from 1905, was established in 1861 and was the property of John Spencer (Coatbridge) Ltd., who also owned the Drumpellier Iron Works. The two works were equipped with mills, puddling furnaces, hammers and forges that could produce up to 4,000 tons of finished iron per month. Their product range included rivet iron, boiler plates, round stanchions, angles and other sections for ship building, iron for bridge, roof and agricultural work, horseshoe iron, wheel tyres and many other items. Much of this output was exported, although a high proportion went to local tube, rivet and other manufacturers. Initially a site beside the canal was regarded as ideal for an iron works, but as the pictures of the Phoenix works show, railways were threaded into every corner of Coatbridge and as attitudes changed some companies, like the Coatbridge Iron Works, one of the town's earliest works, moved away from the canal to another site. The iron industry in all of its branches in Coatbridge was so big and had so many furnaces blazing away, that the town was turned dark with smoke by day and lit up by fires at night, attracting unflattering descriptions like 'no worse place out of hell', or 'hell with the lid off'. The canal, steaming from cooling water being discharged into it, could have doubled as a latter-day River Styx.

To begin with, one bridge carried the main road across the canal at a point where it took a significant bend. The bend will have helped because, in those days, bridge builders set their structures square over whatever they were crossing. Then, in 1826, the Monkland & Kirkintilloch Railway was opened from Palacecraig to the Forth & Clyde Canal passing through the centre of Coatbridge. The canal bridge must have been widened because a level crossing took the railway over the main road and across the canal at the same point. Trains of coal wagons on the railway were initially horse-hauled, so this arrangement would not have seemed odd, it was just another road, one with rails, but as traffic on both road and rails increased and locomotives superseded horses, a railway bridge was built. The road junction became busier and bigger, and the canal bridge evolved into a larger tunnel-like structure with a roof of steel beams that sat just over seven feet above the water and must have been a real ear scraper for horses tramping the towpath. This picture of Coatbridge Cross shows the bridge in the foreground, although the bottom left corner has been partly obscured.

Canal crossings in the centre of Coatbridge proliferated to such an extent that 'the bridges' became a feature of the town as this picture from about 1905 shows. After the enlargement of the main road bridge, the next bridge was a small one that spanned the entrance to the Gartsherrie branch canal, which was cut in 1828 and formed a junction on the offside of the main line, just to the west of the main bridge. A railway bridge, from which this picture will have been taken, was built in 1848 (see page 21). The lattice girder bridge that spans the main road and the canal, and which features prominently here, was built in 1872 for the railway line that had previously run across the road bridge. The gently arched small bridge in front was made for horses hauling barges along the Gartsherrie branch. Before it was made, the horses had to cross the canal by walking up and over the main road bridge, mingling with road and rail traffic, and pedestrians – all potentially dangerous. After the canal had been closed and filled in, the bridges formed part of a town centre landscape feature.

The bridges form the backdrop to this splendid picture of canal activity. On the towpath, on the extreme right, a horse is tucking in to a well-earned nosebag of feed, while the men on the boat take a breather, or in the case of the man standing at the tiller, a moment to smoke his clay pipe. The boat is very simple, a scow with no hatch sides, clearly intended for work only on the calm water of the canal. Scow is a Scots word, which means flat-bottomed boat and was used mainly to describe the ordinary, unpowered canal barges. The word has also crossed the Atlantic to America and is used, with a different spelling, in Holland and other North European countries. The scow is piled high with coal and across the top of the cargo lie two punt poles. The boatmen will have used these to fend off at bridges and other structures, or to re-float the boat if she got stuck anywhere. The plate girder bridge at the entrance to the Gartsherrie branch canal can be seen behind the cabin lum.

In this picture of Sunnyside, beyond the foreground trees and below the embankment and retaining wall, is the Gartsherrie branch, or to give it its full name the Gartsherrie, Hornock and Summerlee Branch. To make the branch canal, the Gartsherrie Burn was diverted and the channel cut along its former course. Initially intended to gain access to coal mines, the new bit of canal just happened to be in the right place for two huge iron works to be built alongside, following development of the hot blast smelting process in 1830. Running along on top of the embankment are the former tracks of the Monkland & Kirkintilloch Railway, which had become part of the Monklands Railway Company in 1848, and which in turn was absorbed by the North British Railway in 1865. On the right of the picture, is the former Coatbridge Central Station, opened at the same time that the railway bridge over the canal was built. Somewhat hazy on the hillside is the Academy building with Gartsherrie Church, at the head of Baird Street, to its left.

In 1828 James Beaumont Neilson developed a process of burning furnace gases to heat air before blasting it into a furnace. Allied to the earlier discovery of blackband ironstone and combined with local splint coal, the Monklands were poised for a dramatic transformation, and the men who led the charge were brothers, William and James Baird. Blatantly flouting Neilson's patent, they set up the Gartsherrie ironworks at the end of the newly-completed branch canal. Neilson, a former business associate, sued; the Bairds paid and carried on smelting. They went on to establish ironworks elsewhere and made a fortune, but Coatbridge was their heartland. They built rows of humble cottages for their workers, but also bought up land and laid out plans for its development: Sunnyside was one of these areas and its crowning glory, Gartsherrie Church, was largely paid for by the Bairds. Gartsherrie Ironworks, seen here about 1910, was the first hot-blast plant in the town, it was also the largest and the last, eventually closing in 1967.

Gartsherrie started a trend. The old cold-blast works at Calder were upgraded and new works were established beside branch canals at Dundyvan, Langloan and Summerlee. Almost a next-door neighbour to Gartsherrie, Summerlee was set up in 1836 by James Beaumont Neilson's brother, John. A number of other family members were involved in the business, which became known as the Summerlee Iron Company in 1870. The works is seen here with the branch canal on the right and the Gartsherrie Burn just to the left of centre. At its height eight furnaces were in operation, but the works closed in 1930. The giant structures were demolished and the spoil used to level the site. It was used by Lambert & Co. Ltd. to make Hydrocon Cranes, but when that closed Summerlee came back to life. The overburden was removed and archaeological excavation in the 1980s revealed the bases of some furnaces. These formed part of Summerlee Heritage Park, which became known as Scotland's noisiest museum. It has been since re-launched as Summerlee, Museum of Scottish Industrial Life.

The Monkland & Kirkintilloch Railway (M & K) represented an early challenge to the canal's coal trade, but just at the moment when the iron industry promised to add significant new traffic, a much more serious rival appeared: the Garnkirk and Glasgow Railway. The new railway was laid to the north of the canal, but on a broadly parallel line, branching off from the M & K near Gartsherrie and terminating at Townhead. It was opened in 1831 with two locomotive-hauled trains, one carrying dignitaries out from Glasgow and the other heading into the city with coal. This engraving, used in the 1850s, shows two such trains passing the Garnkirk colliery and brick works – after coal, bricks were one of the most important commodities carried on the railway. Initially Garnkirk & Glasgow trains used M & K rails for about a mile and a half, but this arrangement ended in 1843 when a new line was laid on the west side of the canal branch between Gartsherrie and the centre of the growing town. At the same time, the company name was changed to the Glasgow, Garnkirk and Coatbridge Railway.

When the Garnkirk and Glasgow Railway built its branch into Coatbridge to bypass the Monkland & Kirkintilloch route, it didn't stop at the canal, but carried on to Whifflet where it joined another early railway, the Wishaw and Coltness. This latter line had been opened in 1834 as an extension of the Monkland & Kirkintilloch Railway from Dundyvan to Holytown and ten years later had been extended south to the Coltness Iron Works at Newmains. This was the great age of railway expansion and in 1848 these local lines became links in one of Scotland's, and Britain's, major through routes when the Caledonian Railway absorbed them as part of their main line north from Carlisle. The 'Caley' also gained a route into the north side of the city along the Glasgow, Garnkirk and Coatbridge line. This railway expansion added two more bridges over the Monkland Canal in central Coatbridge; the railway bridge seen in this picture looking toward the town centre and the footbridge below the main structure. Threaded through arched apertures in the piers of the larger bridge, it is an ingenious piece of Victorian engineering.

Situated to the west of 'the Bridges', Merrystone (or Marystone) Bridge provided a route across the canal from Bank Street to Blairhill Street and is seen here in a picture from about 1910. The canal looks in good working order, with well-tended banks, yet some six decades later it had been culverted and filled in. The bridge survived the culverting process and was later included in a landscaping scheme that was awarded a 'Regeneration of Scotland Design Award'. In the 1950s and 60s, when all this work was instigated, water in the urban setting was regarded as dangerous, a source of disease and a breeding ground for vermin and insects – the fearsome local mosquitos were known as 'flying tigers'. Opinion has since shifted and the modern view is that properly managed water in the built-up environment is a positive asset. It does seem a pity therefore that a lot of money was spent laying twin concrete pipes along the bed of the canal, providing inspection man holes, filling in what was left and then landscaping it, that the end result was a path where previously there had been a path beside a waterway.

More subsidiary canals branched off the Monkland in Coatbridge than off all of the other Scottish canals put together. There was Dixon's Cut, a branch to the Dundyvan collieries and ironworks, the Gartsherrie branch and the Langloan branch, the subject of this picture taken about 1903. A bridge, which carried the towpath over the end of the branch, can be seen with the houses of King Street and Blairhill in the background. Behind the camera, the branch passed under Bank Street and Buchanan Street, through one of the few canal tunnels in Scotland, before terminating in a basin beside the Langloan Ironworks. The works was set up in 1841 by a partnership of Robert Addie, Robert Miller and Patrick Rankin. Miller and Rankin had both left the partnership by 1860 leaving Robert Addie in sole ownership of an extensive iron and coal enterprise that included Rosehall Colliery, one of the largest in the Coatbridge area, but also one associated with some of the worst miners' housing in the country. Langloan Ironworks closed in the depression that set in immediately after the First World War.

A few factory chimneys poke above the distant houses in this picture, taken in 1952. The canal occupies the foreground and although this stretch looks to be in good condition, the waterway had officially been abandoned two years earlier. Some children, who had probably been told by their parents not to go near the canal, are wandering along the banks, with behind them the wide-open spaces of West End Park. Bounded by the canal and the Langloan Branch to the north and east, and to the south and west by Bank Street and Blair Road, the park was gifted to the town by the owner of the Drumpellier Estates. In all probability this simply recognised reality because the Greenmuirs, as it was known at the start of the 19th century, had by the 1850s been partly cleared of trees and was known as Yeomanry Park – the place where local military volunteers practised and paraded. Cricket and other amusements had also become popular by the time it was handed over and it has since been used to stage fun fairs, circuses and, as the picture shows, games of bowls.

After 'the bridges' in the centre of Coatbridge, the canal crossing that features most in old pictures is Blair Bridge. As this picture shows, it was in close proximity to the more upmarket housing of King Street and Blairhill, a locality where postcard publishers were more likely to sell their wares. Industry, however, was never far away in Coatbridge and there were elements of the Drumpellier Colliery nearby, including an old pit sunk beside the bridge. Since this picture was taken, Blair Road has been upgraded and the bridge rebuilt with brick parapets replacing the wooden railings. As part of an environmental enhancement scheme in 2011 a decorative steel entrance arch was erected beside the bridge to give access to the section of towpath through Drumpellier Country Park. It is a delightful stretch of canal and still in water because just to the east of Blair Bridge is where the culvert from Sikeside ends and the water spills out of a concrete weir. The turbulence is clearly attractive to fish, because the water just below the culvert attracts fishermen.

After closure, a remnant of canal, just over a mile long, remained in water through Drumpellier Estate, but while this has modern significance, the estate's former owners were crucial to the canal's early history. When the canal was first proposed, its backers included James Buchanan, son of Andrew Buchanan, one of Glasgow's famous 'tobacco lords', former Lord Provost of the city and owner, since 1735, of Drumpellier. James had inherited the estate on his father's death, but following the American War of Independence and collapse of the tobacco trade, the Buchanan fortunes crumbled and he sold Drumpellier to his cousin Andrew Stirling in 1777. He became principal proprietor of the canal, effectively saving it from an early demise, but when his own personal finances crumbled in 1808, David Buchanan, grandson of the tobacco lord, bought the estate back for his branch of the family. Although mainly used for leisure by these wealthy proprietors, the estate did provide some produce and its 'home farm' steadings occupied ground to the south of the canal. The lattice girder bridge in this picture was built adjacent to the steadings to provide access to the land on the north bank.

The bridge at the Drumpellier home farm was a late 19th century addition, spanning the full width of canal and towpath and not constricting the navigation with a narrow bridge hole. It superseded the earlier Garnheath Bridge, seen here, which appears to have fallen into disrepair and been replaced with a private footbridge by or before 1904. The date corresponds with the last years of Sir David Carrick Robert Carrick-Buchanan, a devoted Lanarkshire man, Deputy Lieutenant of the county, County Councillor, Justice of the Peace, soldier and cricketer. Following his death in 1904, the family showed little interest in Drumpellier and in 1919 gifted the estate to Coatbridge. The mansion house was demolished in 1960 and in 1984 Monklands District Council converted the estate into Drumpellier Country Park. The canal, which forms an important part of the park, was dredged from Blair Bridge to a culvert installed to replace the railway bridge on the line east from Glasgow Queen Street (Low Level) to Coatbridge Sunnyside and beyond. A tiny rump of canal was left to the west of the railway, but from there the water flows through a pipeline all the way to Port Dundas in Glasgow.

Of the canal's many branches, the most remarkable wasn't a waterway, but a railway: the Drumpellier Railway. Constructed between 1847 and 1854, a time when railways were being built as independent transport undertakings all over the country, this two-miles-long line was made as a feeder to the canal. Operated by inclines and a stationary engine, it linked pits to the south of the canal to a loading point at Cuilhill, a pit village to the west of Drumpellier. A large coal trans-shipment operation beside the towpath would have impeded other canal movements, which is presumably why a bridge carried the railway tracks over the main channel onto an island, where they fanned out to lie alongside the wharf edges. The high island and the narrow channels around it became known as the Cuilhill Gullet (a gullet in old Scots can mean a gully). The facility fell into disuse and decay when the railway ceased working in 1896 and when the canal was closed and dewatered the remains were left high and dry, resembling an archaeological site. The function of iron hooks that were left hanging from the wharf walls can only be guessed at.

NOTICE.

A SPECIAL GENERAL MEETING of the MEMBERS of the

CUILHILL CO-OPERATIVE SOCIETY, LIMITED,

WILL BE HELD WITHIN THE

OFFICE OF HENRY M'LACHLAN, ACCOUNTANT, COATBRIDGE,

ON MONDAY, THE 16TH DAY OF DEC. CURT.,

AT FOUR O'CLOCK P.M.,

For the purpose of considering the resolution of the Creditors to wind up the Society by Liquidation, and if so resolved to approve of winding up the same by liquidation.

1 ACADEMY STREET,
COATBRIDGE, *9th December, 1878.*

ALEX. PETTIGREW, PRINTER, COATBRIDGE.

The Drumpellier railway drew coal from pits in the extensive Bredisholm coalfield and other collieries including Braehead, Bargeddie, Drumpark and Rosebank. A separate wagon-way further west took coal from another pit to the Cuilhill Gullet for transfer to canal barge. Cuilhill itself was a small mining village with a pit in the immediate vicinity and others nearby. It was mostly situated on the north bank of the canal and in the mid 19th century amounted to little more than four rows of cottages. Some expansion appears to have taken place in the following years, but while never big, it does appear to have been large enough to support its own retail co-operative society, as this winding-up notice implies. Co-operation was not new when the Rochdale Equitable Pioneers Society was formed in 1844, but the model they established was adopted by co-operative societies in industrial communities, many of which were in mining villages like Cuilhill. The demise of the society could indicate a similar struggle for survival for the pits, the canal and the Drumpellier Railway.

Although the canal to the west of Drumpellier was piped, filled in and consigned to history, a few structural remnants survived, one of which, Netherhouse Bridge, is seen here in 1990. With the old timber deck replaced by brick and concrete, and the railings with steel mesh, it was not pretty, but situated on a minor road it remained in situ slowly succumbing to the ravages of time and vandalism. Beyond the bridge, Bargeddie Parish Church, occupies an isolated site, but one that was handy for many parishioners in the days when people walked to the kirk. The spire, slender and distinctive was one of the most recognisable landmarks along the canal, although by 1876, when the church was built, traffic was dwindling. Conversely, the mining industry was expanding and at Lochwood Pit, just over half a mile to the north of Netherwood Bridge, a pioneering coal-cutting machine, known as the 'Gartsherrie', was tried out in 1864. Plagued by problems, it was abandoned, but the basic design was evidently ahead of its time because, after development in America, it became the standard for the industry in the first half of the 20th century.

Swinton, Easterhouse and West Maryston formed a cluster of little villages around Easterhouse Bridge. They were mainly mining villages, although West Maryston, seen here in a view looking east along the canal, appears to be older than the others and may also have been a weaving village. An empty scow sits on the offside of the canal, which looks tidy and well maintained, with neatly tailored banks and weed-free water, the opposite of the descriptions that sealed its fate in the 1950s and 60s. The name West Maryston implies the existence of another Maryston to the east, which could be Maryston, or Merryston, in Coatbridge. The village was also known locally as the 'Hole', which doesn't sound very flattering, but the name appears on old maps evidently predating West Maryston. Whatever the name or its meaning, people were emotionally attached to their village and its attractive countryside setting, and so when it was demolished after the First World War there was sadness that, instead of the new housing being built in the immediate locality, it was in Baillieston.

The photographer who took this picture was standing close to the spot that the picture on the previous page was taken from, although this view looks west. The towpath edge, as in the previous picture, looks well tended and on it a diminutive (and somewhat blurred) horse is hauling a boat toward camera. The boat, a scow, looks huge compared to the horse, because it is empty and riding high in the water. This was the normal way of working on the canal, for boats going west to be heavily loaded with coal or iron and to be empty when returning to the Monklands. In the background is Easterhouse Bridge with the local school on the right, behind the fence. Going north (right) from the bridge was the Easter House, while to the south were Easterhouse and Swinton, mining villages set in an otherwise rural landscape. The collieries around here were not the super-pits of the 1960s, but relatively small operations working close to the surface in Lanarkshire's rich seams known by names like Kiltongue, Lower Drumgray and Splint.

This evocative view looks east along the canal toward Easterhouse Bridge with a heavily loaded barge passing under it and heading toward camera. It's a pity that the photographer didn't wait for the vessel to come closer, but given that it would have been impossible to stop the unpowered boat, it was probably wiser to get out of the way. An empty scow sits on the offside, beside a shed that was part of stabling where canal boatmen, who lived in the nearby villages, kept their horses. The little West Maryston Public School, facing camera, served the community from the late 19th century, until the 1920s when a new primary school was built in Swinton. Of the three villages, Easterhouse was the youngest, but its name was the one adopted for the huge new housing scheme that transformed the whole area. The canal was also transformed; between Easterhouse and Glasgow Townhead the route of the canal became the route of the M8 motorway and a bridge spanning six lanes of highway and two hard shoulders replaced the little Easterhouse Bridge.

The canal was comprehensively destroyed by motorway construction, although as these pictures show, the dewatered bed proved useful to the contractors as a construction road. Regrettably, the details of when and where the pictures were taken was not properly recorded (by the author!), and so an approximate date of the late 1970s will have to suffice, as will a vague memory that the trucks were passing through Queenslie Bridge. When first built, the section of M8 between Townhead and Easterhouse was known as the Monkland Motorway. It was opened in three sections, from Townhead to the Cumbernauld Road in 1975, to Stepps Road in 1979 and the final section to the Baillieston Interchange in 1980. The road closely followed the line of the canal, so much so that, had the 18th century entrepreneurs and engineers they been able to witness its construction, they might have been quietly pleased that their route was still being used as Glasgow's main eastern transport artery. Bridges over the canal were replaced by road or foot crossings and Queenslie Bridge, if indeed the pictures were taken there, was rebuilt as a footbridge.

A sweeping bend on the motorway replicates this gently curving section of canal, photographed near Bartiebeith Farm in 1906. At the time Easterhouse was noted for its sweet fresh air, which could be enjoyed by people who took a train to Easterhouse Station and walked along the towpath. A contemporary writer, who did just that, described this location as a 'delightful stretch of wood and water … with fields ripening to an abundant harvest, and rising beyond to copses and uplands'. He also wrote about 'the scent of hawthorn, which flourishes on the banks of the canal at this season (summer) in an abundance seldom met with'. That rural idyll has long gone, but the canal's natural world might have left another legacy in Easterhouse. In 2015 scientists announced that they had identified a large colony of water voles living in and around the housing scheme, miles from any obvious source of water. Voles were common on the canal and, although disputed by the scientists, there is something thrilling about the idea of these timid little creatures, deprived of their habitat and frightened by the kind of activity seen on the facing page, going into hiding and quietly thriving until they were discovered years later.

Sharing the canal banks with scented hawthorn and scurrying voles were loading points for boats. This jetty, with a steam lighter and scow alongside, was associated with the Provan Hall Colliery and Fireclay Works. The Provan Hall name was taken from a delightful little mansion house that has similar features to Glasgow's oldest house, Provands Lordship and, like it, could date from the late 15th century. In the background is Bartiebeith Bridge, known locally as the 'Ruffian Brig', a shortening of Bully Ruffian, the name sailors used for HMS *Bellerophon*, the ship that captured Napoleon in 1815. In the 1920s, Glasgow attempted to move the city boundary east to take in this largely rural part of Lanarkshire. It was rebuffed, but by 1938 the city had succeeded in its ambitions. The outbreak of the Second World War thwarted any development aspirations, but when hostilities ceased and faced with an urgent need to rehouse people from older run-down areas, the city began to spread eastwards. Provan Hall, which had been presented to the National Trust for Scotland in 1935, was surrounded by new housing and later incorporated into the new Auchinlea Park.

Garthamlock Colliery's number 5 and 6 pits were situated on the north bank of the canal to the west of Queenslie Bridge. The colliery was part of a small complex of pits operated latterly by the Steel Company of Scotland, whose Blochairn Steel Works was also situated on the north bank of the canal, but three miles to the west. The most southerly pit in the group was Queenslie Colliery and from there a continuous haulage mineral railway ran north to the Garthamlock pits, crossing the canal on the bridge on the right of the picture. Beyond its connection with the Garthamlock pits, the railway carried on up the hill to a large screening plant adjacent to Comedie Pit and, from there, across Cardowan Moss to join the main line tracks of the former Garnkirk and Glasgow Railway. The barge in the picture suggests that some coal was sent to the steelworks by canal, but the railway infrastructure is an indication that it was not the favoured option by the early 20th century. The pits had ceased to operate by about 1930.

Used as a postcard in 1906, this picture was sent to a Shettleston address by 'Dad'. In his message he described the scene as, 'cottage by the canal and Teddy, but you can't see Teddy because he has gone inside to buy chocolates'. The inference from this comment is that the cottage, a canal employee's dwelling situated beside Milncroft Bridge, doubled as a small shop. The view looks east with what appears to be the Cranhill Fireclay Works in the distance. It made sense to move heavy products like bricks and tiles by boat, which partly explains the number of fireclay works beside the canal, but they were also there because large quantities of clay were dug out of local collieries. Coal is the fossilised remains of long dead trees, fireclay the solidified mud they grew in – and so brick and tile works often operated alongside coal mines. Both the cottage and Cranhill Fireclay Works had been demolished long before motorway construction utterly altered this scene. Spanning the new road was a footbridge, built where the old Milncroft Bridge had been to provide a link between Ruchazie in the north and Cranhill to the south.

Situated on the north bank of the canal beside the Gartcraig or Jessie's Bridge, the Gartcraig Fireclay Works was set up in 1872 and remained in operation until 1921. It is seen in the distance of this view looking east along the canal, with the chimneys of the Gartcraig Rows just visible above the hillside on the left. Clay extracted from the Gartcraig Colliery and used for brickmaking came from a seam below the 'Balmoral' coal. The seams in this part of Lanarkshire were not flat, they undulated, varied in thickness and were often disturbed by faults and intrusions. Consequently, in the days before scientific testing, knowing exactly where in the geological strata a pit was situated could be a hit and miss process reliant on practice and experience, and at Gartcraig there were doubts as to exactly what seams they were working. Managers thought that the coal they called the Virtuewell was in fact the Kiltongue seam and the one being worked as Kiltongue was in reality the Balmoral. As a seam it was rarely as much as two feet thick, which would have made it difficult to mine, but the underlying fireclay made it more workable and thus profitable.

There was another fireclay works beside the canal at Barlinnie, but a new landmark appeared in 1882 when 'A' Hall, the first phase of Barlinnie Prison was opened. It could hold up to 200 prisoners and when Halls 'B' to 'E' were completed by 1897, the prison's capacity reached 1,000. It was built to ease pressure on the notorious Duke Street Prison, although that grim building remained in use, latterly as a women's prison until 1955. Barlinnie, also colloquially known as the Bar-L, or the Riddrie Hilton, is seen here depicted, curiously, on a postcard – an item more usually associated with saucy seaside fun, great buildings or scenic splendour. The view, which looks east from Smithycroft Bridge, shows the canal in the early 20th century looking well maintained and in good order. It also shows it at a point close to where James Watt terminated the first phase of construction in 1773. The money had run out, but near a road and not far from the city, Watt believed that the canal would be 'of immediate and profitable use because even from that termination we can afford to undersell others'. He was wrong.

Rural crafts, swept away by the city's eastwards march, are reflected in the names of these canal bridges: Milncroft could have been derived from a miller's croft and Smithycroft from a blacksmith's. At the time that Barlinnie Prison was built, Smithy Croft, seen here, was a village beside the canal bridge, out of picture to the right, and Riddrie was a farm, but after the First World War things began to change. A major pre-war report into the condition of workers' housing in Scotland highlighted huge problems, but little could be done at that time. Then, after the war, local authorities with no prior experience were given responsibility for housing provision, previously the preserve of private developers. Glasgow Corporation's first step into this brave new world, in 1920, was a housing scheme at Riddrie, bounded to the north and west by the Monkland Canal. As housing spread, so did the associated infrastructure and Cumbernauld Road was realigned for trams, with a new Smithycroft Bridge being built by 1925, although Lanarkshire County Council were unimpressed when Glasgow attempted to enlist their help to pay for it.

Following the development of Riddrie, housing spread east along the Monkland Canal, transforming the landscape from rural to urban. One of those developments, Blackhill is on the left of this grainy, mid-winter picture from the late 1950s, with Riddrie on the right. A concrete footbridge connects the two areas of housing across the partly dewatered, semi-derelict canal to the west of Smithycroft. The canal at this location was made during a second phase of construction, which began in 1780 and continued until 1783. It consisted of two separate sections of canal, one extending from Watt's terminal near Barlinnie, to the top of Blackhill and another from the base of the hill to Glasgow Townhead. The expense of constructing locks was avoided by connecting the two sections of canal with a dry inclined plane, with trucks that could be lowered down or hauled up the hill on rails. Boats on the upper and lower levels had to be unloaded and reloaded, which could have been costly, because people at that time liked to see what they were getting when buying coal, so preferred it in large lumps – 'big coal' – and the double-handling probably broke up the lumps into smaller bits.

Although the second phase of construction had brought the canal to a terminal at Townhead it remained unprofitable, due in part to the cumbersome Blackhill haulage system. The problem of what to do about the hill therefore had to be solved when the decision was made to link up with the Forth & Clyde Canal, and so, when work began in 1790 on the third phase of construction, one of the principal tasks was to build a flight of locks at Blackhill. The height difference between the upper and lower levels was 96 feet, so this was a prodigious task. Initially eight locks were built in four staircase pairs, each pair being made as a single structure with a common middle gate. Elongated basins between the locks acted as reservoirs and allowed ascending and descending boats to pass. The lock flight performed adequately for some 40 years, but the rapid growth of traffic after 1830 meant that a second flight of locks was needed. Completed in 1841, it was built alongside the first, in effect duplicating each pair as a group of four and making a total of sixteen locks. The four locks at the top of the flight are seen here in a picture taken around 1900.

Glasgow industry's insatiable appetite for the products of the Monklands placed enormous pressure on what became a bottleneck at Blackhill. The initial response, to double up the lock flight failed to cope with the increasing volume of traffic and made demands on water supplies that at times proved too great. Something else had to be done and an idea that had been rejected when the locks were doubled was resurrected. The proposal was to install an inclined plane fitted with rails and a haulage system capable of moving entire boats. It was built in 1850 to the designs of engineer James Leslie, who took advantage of the fact that boats going east were usually empty. This allowed him to design tanks that were just deep enough to keep the boat supported in water, and to mount these on wheeled cradles that kept the boat level as it was hauled up the hill by cable. A steam engine at the top of the incline provided the motive power. The tanks were known as caissons, a word Glaswegians appear to have 'modified' to 'gazoon' when referring to the inclined plane. Although this picture was taken from Alexandra Park many years after the plane was abandoned, its sloping track and haulage engine chimney can be seen to the right of the locks.

Through the middle decades of the 19th century a constant stream of boats went west with coal and iron, while any cargoes going the other way were usually of ironstone, limestone and (and this really wasn't glamorous) manure – animal and human, to be spread on fields between Blackhill and Drumpellier. Boats could take an hour to go through the locks, but less than ten minutes on the plane and with initially heavy usage the haulage engine could be kept in steam, but it became uneconomic to do that as traffic diminished in the 1880s and the plane fell into disuse. The locks remained operational and are seen here in a picture from around 1910. The railings in the foreground are guarding the edges of the towpath, which ran along a causeway between the main canal and the side arm leading to the foot of the inclined plane. A small boatyard was situated alongside the side canal that led to the top of the plane; an earlier boatyard at the foot of the locks was moved to above the locks when they were doubled and may have been moved again to accommodate the plane. It was mainly a repair and maintenance yard, although some boat building was carried out. It may also have been where people joined a boat for a trip into the country (see page 1).

At the start of the 20th century the canal's iron trade was finished and although new power stations at Port Dundas on the Forth & Clyde Canal created another market for coal, traffic continued to dwindle. Boat movements had ceased by the mid 1930s and the deserted canal languished, growing weed and silting up until the Second World War. Anxious to cut costs, the London, Midland and Scottish Railway sought permission to abandon the canal, but water supplies to the power stations and the Forth & Clyde remained important and so the government refused. Unwanted and thus neglected the old waterway sank rapidly into dereliction and became an object of concern. Throughout its existence the canal had been the scene of drowning tragedies, some accidental, some self-inflicted, but in its abandoned state and serving no evident purpose, this loss of life became the focus of public disquiet, with much of the criticism directed at the decaying Blackhill locks. The canal was formally abandoned in 1950, but not piped and filled in before this picture was taken about 1957 of what looks like a steaming, rubble-strewn swamp below the Blackhill Locks.

Coatbridge wasn't alone in having a malleable iron industry; one of the largest concerns in Scotland was situated beside the canal at Blochairn, in Glasgow. When it failed, the works, furnaces and other plant lay moribund for about six years until the Steel Company of Scotland bought it all for a knockdown price in 1880. Set up in 1872 the company operated a large plant at Hallside, Cambuslang and had intended to move the Blochairn machinery there, but changed their minds and kept their new acquisition working. Steel was the new thing and the Steel Company were innovators, who pushed the boundaries, but it wasn't a commercially sound strategy and in the early 20th century, lost their pre-eminent position to the upstart, David Colville and Sons. In the following decades Scotland's steel industry went through a process of rationalisation with the Steel Company and others being absorbed into Colville's. The Blochairn works remained active through the Second World War, but with some of its plant modernised and other elements obsolete, it closed in 1952. The picture shows Blochairn workers in 1915.

In the second phase of construction, between 1780 and 1783, the canal was extended from Blackhill to Townhead where a small terminal basin was created. From there, in the third period of construction, the canal headed north through Garngadhill Bridge and continued for about 200 yards before turning sharply west again to pass under the Castle Street Bridge and join the Forth & Clyde Canal. That short north-south section of canal is seen in this picture from the 1950s, with the coal wharf on the left and the soot blackened tenements of Earlston Avenue on the right. It may have been atmospheric, but Glasgow's post-war planners wanted to erase images like this and homed in on this spot in 1965 to start construction of the hugely complicated Townhead Interchange, the first phase of the Inner Ring Road motorway. The ring was never completed, so the junction's designed role was not fulfilled, but a connection was made with the Springburn Expressway, which cut through the St Rollox Basin of the Forth & Clyde. Remarkably, amidst all the demolition and reconstruction, one small remnant of the Monkland Canal survived: the Castle Street Bridge.